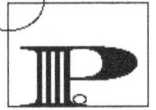

ENERGY OF THE STATE 2008

Published by Fast Entropy Press,
a division of the Pavilion of Research and Commerce
San Francisco, California, USA

www.energyofthestate.com

www.fastentropy.com

Enquiries should be sent to fastentropy.pavilionrc@earthlink.net

Abstract

Physical data are collected regarding population, energy production, energy consumption and reserves for the member countries of the North American Free Trade Agreement (NAFTA). Data are then analyzed to develop a snapshot and modest projections for each of the member countries and for the NAFTA area as a region. Additional anecdotal sources are then consulted to identify potential sources of change.

1. INTRODUCTION

The Energy of the State report attempts to provide a snapshot of physical energy production and consumption of each of the large North American countries. Simple projections are provided as well. Those countries are Canada, Mexico and the United States of America, which are also the three members of the North American Free Trade Agreement (NAFTA). Physical quantities are considered rather than financial data. Since the report is presented on a nation-by-nation basis, it is titled "Energy of the State." Since for physical purposes, the NAFTA area can be considered as an entire region, regional and per capita figures are included for the NAFTA area.

This report is experimental in nature and it is not comprehensive. The figures states are approximate, and may even be incorrect in some instances. The figures stated also involve significant calculations in which rounding errors may have occurred. Rather than polish the numbers by fixing the errors, the figures have been left in rough form to emphasize their rough nature. Consequently, not all of the totals are the exact sum of their constituent figures.

2. METHODOLOGY

2.1 Sample Selection

The three members of the North American Free Trade Area (NAFTA) were considered. These countries are Canada, Mexico and the United States and are by far the largest economies in North America. There are other territories in North America such as Greenland, but they are not presently a part of NAFTA. Energy production and consumption of the non-NAFTA North American territories is relatively small.

2.2 Sources of Data

The primary source of energy data was the United States Department of Energy's Energy Information Administration (EIA). Other sources were used as noted. While the data and calculations ultimately behind U.S. government figures have been criticized (Deffayes, 2001) and perhaps rightfully so, the EIA is still regarded as the most commonly used and accepted source. However, the methods used by this study to total EIA figures are in some cases different than that used by the EIA, and units reported by the EIA have nearly always been converted.

Industry reports were used to supplement, verify and critique EIA statistics, particularly regarding uranium production and reserves. The validity of these reports has only been verified in terms of reasonableness, not accuracy. Population figures were obtained from the US Census Bureau for the United States, and BBC Country Profiles (citing United Nations figures) for Canada and Mexico.

EIA statistics do not go up to 2008. In fact, even for the immediately preceding few years (e.g. 2005 or 2006), only projections are provided. Instead of using projections to make further projections, the latest figure or projection was utilized. In that sense, this report could be dated 2006 or even earlier. However, year-to-year differences are not typically great, so for purposes of a snapshot, these figures are reasonably representative of nearby years.

2.3 Utilization and Critique of Data

The data was typically accepted as stated. This may not be entirely accurate, but no entirely reliable different figures were located. These stated figures were either considered to be the best available, or were at least the most widely cited. Where multiple sources were considered, the most moderate or credible figure was selected, and any extraordinary differences were noted.

Only the largest sources of energy were considered in the main calculations, specifically petroleum, natural gas, coal, nuclear and hydroelectric. The totals for each country and NAFTA are calculated from these figures only. This means that the totals may under-report total energy production and consumption by a few percent due to neglecting minor sources.

Renewable sources and their quantities have been identified separately for the United States. Sample calculations of totals for NAFTA and member countries have been provided for both the cases of with and without non-hydroelectric renewable sources to illustrate the magnitude of error by neglecting non-hydroelectric renewable sources and the paradox created by including them.

Data was organized first by country, then by energy source. Figures were converted into Peta Joules (1 quadrillion Joules or PJ) for absolute figures. The figures were further converted into kilowatt hours designated as "kwh". Although Joules are a more elegant unit than kilowatt hours, more North American consumers are familiar with kilowatt hours than Joules. Therefore, the significance of per capita figures reported in this kilowatt hours are easier to appreciate. Joules and kilowatts are both units of energy. Peta Joules means 1 quadrillion Joules. There are 1055 Joules in a British Thermal Unit (Btu). 1 watt is 1 Joule/second.

Totaling diverse energy sources might appear to be mixing apples and oranges. This criticism is valid for purposes such as powering aircraft, and to a lesser extent, automobiles. It is reasonable to ask whether such a quantity is meaningful. For purposes of generating heat and electricity, totaling diverse sources is much more reasonable. For purposes of calculating heat released into the environment, such a total is essentially the definition of such.

The national totals calculated in this study deviate significantly from the totals reported by the EIA. There are several reasons for this. The largest source of difference is the handling of nuclear energy data for the production of electricity. This study (EoS) considers nuclear energy to be produced in the country where it was mined, not where the reactor was located. This difference is illustrated in the nuclear energy production versus consumption figures. For example, the United States generates a large percentage of its electricity using nuclear reactors, but it imports much of its nuclear fuel from other countries. The generation of that electric power in the United States is reflected under US nuclear consumption. The situation is reversed for Canada.

Similarly, another source of difference between the EoS versus EIA totals may be uncounted movements of energy between countries. For example, if a truck purchases gasoline in Mexico, but carries out most of its transport in the United States, it is possible that that such energy would be improperly reported under Mexican consumption in one total but not another.

Another significant source of inaccuracy in both the EoS and EIA country totals may be specifically related to the unconsidered movement of hydroelectric energy. For example, a quick review of the data will show that hydroelectric production is equal to hydroelectric consumption in the case of each country, but of course the Northeastern US purchases hydroelectric power from Canada. With power being rerouted across large areas of North America at a moment's notice, it may be difficult to ascertain where a particular unit of power that was produced at a particular location was consumed.

In this regard, the totals for NAFTA may be more accurate than the totals for individual countries. Electric power freely moves across national borders, but so far not in and out of North America. Fuels are another matter. Canada apparently exports a considerable of amount of uranium outside of North America, while the United States imports much of its uranium from non-NAFTA countries such as Australia.

Identifying and calculating nuclear energy consumption was fairly straightforward using EIA figures. However, calculating nuclear energy production was more challenging. The nuclear production figures are based upon production of U_3O_8, a common and relatively safe form of uranium. Figures from the EIA and other sources are typically reported in units of mass and weight such as metric tons and pounds. However, U_3O_8 can be enriched to a wide range of levels and utilized by a wide variety of reactors, each with their own rates of energy production. To overcome this issue, a mean energy production figure was utilized. This figure was calculated by dividing global nuclear electric power production by global U_3O_8 production. This isn't a very precise approach for particular countries, but it was the most straight-forward one. This approach assumes that substantially all reported uranium fuel production results in electricity production. It also assumes that substantially all nuclear fuel comprises U_3O_8. These are not great assumptions, but it is not expected that they will greatly distort the accuracy of the nuclear energy production figures.

Alternative approaches to calculating nuclear energy production were considered. One such approach would be to calculate the number of U235 atoms produced (as a percentage of U238 mined) and then to multiply that quantity by the amount of energy released in the most commonly utilized nuclear reaction. Such approaches are physically satisfying, may not account properly for the variety of reactor techniques used, and so may not properly weight the results.

2.4 Other Sources of Information

News articles and industry reports also were utilized, particularly for renewable energy sources and new technologies. Treatises were utilized at times, particularly researching processes, critiquing figures and verifying conversion factors.

3. DATA

3.1 Organization of Data

The data is organized into several tables. These tables do not show calculations, but standard conversion formulas applied to the publicly reported source data can be used to verify these figures. The first three tables are relatively comprehensive. Additional tables are provided to examine details or provide additional perspective.

3.2 Description and Critique of Tables

Table 1 reports recent population figures for each of the three NAFTA countries. The sources include United Nations data as reported by BBC Country Profiles and a recent U.S. Census press release. Such figures may undercount the actual populations.

Table 2 reports recent total national production, consumption and net figures for NAFTA countries and for the NAFTA area. A surplus of production versus consumption is reported as a positive number, while a deficit of production versus consumption is reported as a negative number. Sources include the EIA and industry reports. Arguably the real figures for US energy consumption should be larger than reported, since the US effectively uses a net amount of energy that is reported to be consumed elsewhere by having a large merchandise trade deficit. This same situation could apply to Canadian and Mexican figures as well.

Table 3 reports energy production, consumption and net (surplus/deficit) on a per capital basis. Sources include the EIA and industry reports.

Table 4 reports existing reserves and then projects how long those reserves would last in terms present *production* rates. Sources include the EIA and industry reports. There are several issues that should be noted at the outset. First, reporting years remaining in terms of present *consumption* would provide startling, thought-provoking projections that could be used by activists for political purposes and by commentators for improved ratings, but that would provide much less accurate projections. This is because consumption is not presently well correlated to production within individual countries, nor may be so in the future. Therefore EoS uses present production rates instead of current consumption rates. Calculations of projections based on present consumption rates are left to the reader's amusement.

Second, the simple calculation used here merely divides present reserves by *present* production. In reality, production will likely not continue exactly at the present production rate and then drop to zero. Rather, a much more gradual decrease is expected, and therefore nonrenewable energy resources will likely last for a longer period, but at a significantly reduced production rate.

Third, the figures reporting reserves are based upon the price paid for that energy source. Some petroleum deposits that are not economical to exploit at $10 per barrel become so at $100 per barrel. Therefore some deposits were not included as reserves in past years of lower energy costs are included now. Therefore, it is possible to have increasing reported reserves in years of increased energy costs, even if there have been no new discoveries. Nevertheless, a doubling of prices will not typically result in a doubling of reserves.

Fourth, even if energy production roughly follows the projections, available energy to a country varies from them. It is possible to buy and sell fuels among countries including non-NAFTA countries. Depending upon the economic and political circumstances of a country, it may be able to procure additional energy or have to sell its energy elsewhere.

Fifth, there are tremendous differences in figures for petroleum reserves for Canada, even as reported by the EIA. The EIA uses several sources, such as *the BP Statistical Review of World Energy* and the *Oil & Gas Journal*, each with different estimates. For example, The *BP* figure is 16.5 billion barrels while the *Journal* figure is 179.21 billion barrels. This is chiefly because the potential contribution of Canadian of sands to petroleum reserves varies tremendously according to what source is used, what they include and their estimates of costs of extraction. Here, a conservative figure is utilized, but this issue throws significant uncertainty into the remaining years of production, very possibly in the direction of understatement.

Table 5 shows hydroelectric production by country and for the NAFTA region. As noted above, both EoS and EIA state consumption by country as identical to production, so that only the NAFTA total provides a particularly accurate total of hydroelectric consumption.

Table 6 presents renewable energy production for the United States. Unlike the case of hydroelectric power, the US production figures are expected to be nearly identical for the US consumption of other categories of renewable sources. The percentage figures shown represent the percentage of *all* energy production, not the percentage of renewable production.

Table 7 illustrates the differences between the EoS and EIA totals, if US non-hydroelectric energy sources (such as geothermal and biomass) are neglected.

Table 8 illustrates the differences between EoS and EIA totals if US non-hydroelectric renewable sources are included. Paradoxically, by including all US renewable energy sources, the US difference is smaller, but the NAFTA difference is larger. This suggests the significance of unconsidered international trade and movements of energy.

Total energy production from the listed sources is not presented in terms of years remaining. Despite the ease of calculation, such a figure would have little meaning and could in fact be misleading. Energy sources are only convertible from one to another at a cost, and often that cost is extremely high. While this study often totals up diverse energy sources as if mixing apples and oranges to make fruit salad, it draws the line at mixing energy sources to make consolidated projections.

Table 1: Population Figures

Country	Population (people)
Canada	32,000,000
Mexico	107,000,000
USA	301,000,000
NAFTA	439,000,000

For sources, see discussion. Figures are intended to be accurate to three significant digits.

[Please continue to the next page.]

Table 2: Production and Consumption (National and Regional Totals)

(All figures are in terms of years unless otherwise noted)

Country	Energy Source	Production (PJ)	Consumption (PJ)	Surplus (PJ)
Canada	Petroleum	7,340.00	5,030.00	2,310.00
	Natural Gas	7,090.00	3,550.00	3,530.00
	Coal	1,510.00	1,350.00	161.00
	Nuclear	8,700.00	1,040.00	7,650.00
	Hydroelectric	3,800.00	3,800.00	0.00
	Total	28,400.00	14,800.00	13,700.00
Mexico	Petroleum	8,280.00	4,460.00	3,820.00
	Natural Gas	1,850.00	2,140.00	-286.00
	Coal	276.00	459.00	-183.00
	Nuclear		109.00	-109.00
	Hydroelectric	290.00	290.00	0.00
	Total	10,700.00	7,460.00	3,240.00
USA	Petroleum	18,700.00	42,600.00	-27,500.00
	Natural Gas	20,100.00	23,600.00	-3,560.00
	Coal	25,300.00	24,300.00	1,030.00
	Nuclear	1,180.00	8,610.00	-7,430.00
	Hydroelectric	2,850.00	2,850.00	0.00
	Total	68,100.00	106,000.00	-37,500.00
NAFTA	Petroleum	34,300.00	55,700.00	-21,400.00
	Natural Gas	29,000.00	29,300.00	-313.00
	Coal	27,100.00	26,100.00	1,010.00
	Nuclear	9,880.00	9,760.00	118.00
	Hydroelectric	6,940.00	6,940.00	0.00
	Total	107,000.00	128,000.00	-20,600.00

Totals do not include non-hydroelectric renewable energy sources.

For sources, see discussion. Figures are intended to be accurate to three significant digits. Figures are displayed to two digits after the decimal point for purposes of visual consistency only.

Table 3: Production and Consumption (Per Capita)

(All figures are in terms of years unless otherwise noted)

Country	Energy Source	Production (kwh)	Consumption (kwh)	Surplus (kwh)
Canada	Petroleum	63,700.00	43,700.00	20,100.00
	Natural Gas	61,500.00	30,800.00	30,700.00
	Coal	13,100.00	11,700.00	1,400.00
	Nuclear	74,500.00	9,040.00	66,400.00
	Hydroelectric	33,000.00	33,000.00	0.00
	Total	247,000.00	128,000.00	119,000.00
Mexico	Petroleum	21,600.00	11,600.00	9,970.00
	Natural Gas	4,840.00	5,590.00	-746.00
	Coal	721.00	1,200.00	-478.00
	Nuclear		285.00	-285.00
	Hydroelectric	757.00	757.00	0.00
	Total	27,900.00	19,500.00	8,460.00
USA	Petroleum	17,300.00	42,600.00	-25,400.00
	Natural Gas	18,500.00	21,800.00	-3,290.00
	Coal	23,400.00	22,400.00	952.00
	Nuclear	1,090.00	7,940.00	-6,850.00
	Hydroelectric	2,630.00	2,630.00	0.00
	Total	62,900.00	97,400.00	-34,600.00
NAFTA	Petroleum	21,700.00	35,200.00	-13,500.00
	Natural Gas	18,300.00	18,500.00	-198.00
	Coal	17,200.00	16,500.00	638.00
	Nuclear	6,240.00	6,170.00	74.60
	Hydroelectric	4,390.00	4,390.00	0.00
	Total	67,800.00	80,800.00	-13,000.00

Totals do not include non-hydroelectric renewable energy sources.

For sources, see discussion. Figures are intended to be accurate to three significant digits. Figures are displayed to two digits after the decimal point for purposes of visual consistency only.

Table 4: Reserves and Projections (National and Regional Totals)

Country	Energy Source	Reserves (PJ)	Production/Yr (PJ)	Years
Canada	Petroleum	101,000.00*	7,340.00	13.70*
	Natural Gas	60,600.00	7,090.00	8.54
	Coal	158,000.00	1,510.00	105.00
	Nuclear	333,000.00	8,700.00	38.10
	Total	653,000.00	24,600.00	
Mexico	Petroleum	83,600.00	8,280.00	10.10
	Natural Gas	15,700.00	1,850.00	8.49
	Coal	29,100.00	276.00	105.00
	Nuclear**			
	Total	129,000.00	10,400.00	
USA	Petroleum	183,000.00	18,700.00	9.79
	Natural Gas	221,000.00	20,100.00	11.00
	Coal	5,840,000.00	25,300.00	230.00
	Nuclear	257,000.00	1,180.00	217.00
	Total	6,500,000.00	65,300.00	
NAFTA	Petroleum	368,000.00	34,300.00	10.70*
	Natural Gas	298,000.00	29,000.00	10.30
	Coal	6,020,000.00	27,100.00	222.00
	Nuclear	590,000.00	9,880.00	59.70
	Total	7,280,000.00	100,000.00	

*See discussion regarding oil sands.
** Mexico has Uranium reserves, but they are not economical to exploit at current prices.

Totals only include the sources listed.

For sources, see discussion. Figures are intended to be accurate to three significant digits. Figures are displayed to two digits after the decimal point for purposes of visual consistency only.

(All below figures are in terms of years unless otherwise noted)

Table 5: Hydroelectric Power Generation

Country	Production (PJ)
Canada	3,800.00
Mexico	290.00
USA	2,850.00
NAFTA	6,940.00

Table 6: USA Renewable Energy Production

Country	Energy Source	Production (PJ)	% of Total Production
USA	Biomass	3,300.00	4.73%
	Hydroelectric	2852.00	3.88%
	Geothermal	361.00	0.49%
	Wind	188.00	0.26%
	Solar	70.10	0.10%
	Total	6,950.00	9.46%
	Total w/o Hydro	4,100.00	

Table 7: EoS versus EIA Production Totals *without* Non-Hydro USA Renewable Sources

Country	EoS Total (PJ)	EIA Total (PJ)	Difference (PJ)
Canada	28,400.00	20,100.00	8,300.00
Mexico	10,700.00	10,800.00	-100.00
USA	68,100.00	73,500.00	-5,400.00
NAFTA	107,200.00	104,400.00	2,800.00

(All below figures are in terms of years unless otherwise noted)

Table 8: EoS versus EIA Production Totals *with* Non-Hydro USA Renewable Sources

Country	EoS Total (PJ)	EIA Total (PJ)	Difference (PJ)
Canada	28,400.00	20,100.00	8,300.00
Mexico	10,700.00	10,800.00	-100.00
USA w/ nh rnw	72,200.00	73,500.00	-1,300.00
NAFTA w/ nh rnw	111,300.00	104,400.00	6,900.00

[Please continue to the next page.]

4. ANALYSIS

4.1 Organization of Analysis

Individual countries are discussed, followed by a summary of data for each country. The NAFTA region is discussed and summarized at well. Then particular energy sources are discussed, followed by a discussion of energy flows. Individual energy consumption is briefly presented. Future technologies and potential trends are presented.

4.2 Country and Region Discussions

Canada, Mexico, the United States and NAFTA are each discussed and summarized in this section.

4.2.1 Country Discussion: Canada

Canada is by far North America's largest consumer of energy on a per capita basis, yet produces enough energy to be a major exporter of energy as well.

Presently, Canada produces a modest surplus of petroleum. That surplus in petroleum is expected to grow due increased exploitation of its oil sands in the face of small increases in domestic demand. (Canada is already economically developed and has fairly low population growth). As mentioned, there are tremendously varying figures for petroleum reserves contained in the Canadian oil sands that could render Canada as either a major or minor player in petroleum production.

Canada presently produces a healthy surplus of natural gas. Yet, Canada's reserves of natural gas are insufficient to meet its projected needs for even a single decade. Yet since natural gas is widely used for heating homes, Canada could have a tremendous dependency upon natural gas due to its cold climate. Natural gas could be Canada's Achilles' heel of energy. Ironically, exploiting oil sands is presently requiring explosive increases in natural gas usage (to produce steam required for the extraction process), potentially leading to shortages.

Canada produces approximately enough coal to meet its own needs. Canada produces a hefty surplus of nuclear power in the form of uranium fuel for nuclear reactors that typically produce electric power. In fact, Canada is the world's largest exporter of uranium although Australia possesses larger reserves. Canada is a significant consumer of nuclear energy on a per capita basis.

Canada produces an amazing amount of hydroelectric power, both in national terms and especially in per capita terms. Hydroelectric power production is equal to about a quarter of Canada's per capita energy consumption.

Canada's medium term energy future appears to be a product of coal, nuclear energy and hydroelectric power. More generous estimates of Canadian oil sands may prolong present levels of petroleum production for perhaps up to two additional decades. If so, petroleum will be a moderately significant part of the medium-term.

Canada's long-term sustainable future will likely be built around hydroelectric power. Biomass could be a sustainable substitute as a home-heating fuel—wood in Canada's case. Due to Canada's latitude, solar power does not appear to be a strong option. Even during the long hours of sunlight in the Canadian summers, the high latitude results in sunlight attacking a low angle with high air mass. Due to Canada's overall low population density, wind can be an important contributor. Similarly, geothermal energy may be significant as well.

[Please continue to the next page.]

CANADA

(All figures are in terms of years unless otherwise noted)

Production and Consumption (National Totals)

Energy Source	Production (PJ)	Consumption (PJ)	Surplus (PJ)
Petroleum	7,340.00	5,030.00	2,310.00
Natural Gas	7,090.00	3,550.00	3,530.00
Coal	1,510.00	1,350.00	161.00
Nuclear	8,700.00	1,040.00	7,650.00
Hydroelectric	3,800.00	3,800.00	0.00
Total	28,400.00	14,800.00	13,700.00

Production and Consumption (Per Capita)

Energy Source	Production (kwh)	Consumption (kwh)	Surplus (kwh)
Petroleum	63,700.00	43,700.00	20,100.00
Natural Gas	61,500.00	30,800.00	30,700.00
Coal	13,100.00	11,700.00	1,400.00
Nuclear	74,500.00	9,040.00	66,400.00
Hydroelectric	33,000.00	33,000.00	0.00
Total	247,000.00	128,000.00	119,000.00

Reserves and Projections (National Totals)

Energy Source	Reserves (PJ)	Production (PJ)	Years
Petroleum	101,000.00*	7,340.00	13.70*
Natural Gas	60,600.00	7,090.00	8.54
Coal	158,000.00	1,510.00	105.00
Nuclear	333,000.00	8,700.00	38.10
Total	653,000.00	24,600.00	

* See discussion regarding Canadian oil sands.
For sources and notes, please see Section 3.

4.2.2 Country Discussion: Mexico

Mexico has the second largest population of the NAFTA members, at over 100 million people. Mexico's economy has both well-developed and developing components. Mexico has an overall surplus of energy production.

Mexico is a significant exporter of petroleum. Although per capita consumption of gasoline is low, Mexico has a large population. It's total consumption of petroleum is over half of its production.

Mexico runs small deficits in other energy sources, including coal, natural gas and nuclear power. Mexico has a single nuclear power plant that provides a small but important portion of Mexico's electric power. Mexico produces a small amount of hydroelectric power.

Mexico faces a mixed energy future. Mexico only has about a decade worth of petroleum and natural gas reserves at present production and price levels. Mexico could face petroleum energy vulnerability in the future, especially if future economic pressures require Mexico to export most of its petroleum production. Mexico possesses over a century of coal reserves at the present production rate. Mexico's medium-term energy future appears to be built around coal.

For purposes of long-term sustainability, Mexico's present energy resources are limited. Nevertheless, Mexico should have a literally bright long-term energy future in terms of solar energy. Due to Mexico's low-mid latitude, much of Mexico receives considerable sunlight without as much cloud cover as equatorial latitudes. The future cost of solar energy will perhaps be the largest factor in Mexico's long-term energy future.

[Please continue to the next page.]

MEXICO

(All figures are in terms of years unless otherwise noted)

Production and Consumption (National Totals)

Energy Source	Production (PJ)	Consumption (PJ)	Surplus (PJ)
Petroleum	8,280.00	4,460.00	3,820.00
Natural Gas	1,850.00	2,140.00	-286.00
Coal	276.00	459.00	-183.00
Nuclear		109.00	-109.00
Hydroelectric	290.00	290.00	0.00
Total	10,700.00	7,460.00	3,240.00

Production and Consumption (Per Capita)

Energy Source	Production (kwh)	Consumption (kwh)	Surplus (kwh)
Petroleum	21,600.00	11,600.00	9,970.00
Natural Gas	4,840.00	5,590.00	-746.00
Coal	721.00	1,200.00	-478.00
Nuclear		285.00	-285.00
Hydroelectric	757.00	757.00	0.00
Total	27,900.00	19,500.00	8,460.00

Reserves and Projections (National Totals)

Energy Source	Reserves (PJ)	Production (PJ)	Years
Petroleum	83,600.00	8,280.00	10.10
Natural Gas	15,700.00	1,850.00	8.49
Coal	29,100.00	276.00	105.00
Nuclear*			
Total	129,000.00	10,400.00	

*See discussion regarding Mexican uranium reserves.
For sources and notes, please see Section 3.

4.2.3 Country Discussion: United States

With a population of over 300 million people, the United States has the largest population of any of the NAFTA members, with over 300 million people. With its large population and perhaps the most highly developed economy in the world, the US is also presently the largest consumer of energy in the world in terms of total usage, but not per capita usage. The US has offloaded some of its need for energy to power manufacturing by importing an increasing percentage of its manufactured goods from abroad.

The United States uses much more energy than it produces in all areas except for coal. Although the United States is one of the largest producers of petroleum in the world, it is also one of the largest importers of it. The US is a tremendous importer of petroleum and is highly dependent on non-NAFTA supplies. The US imports over half of its petroleum consumption.

The situation regarding natural gas is considerably better. Although the US is a large importer of natural gas, its imports constitute less than a quarter of consumption. The US produces slightly more coal than it consumes.

The United States is only a small producer of nuclear energy despite possessing considerable uranium reserves. In contrast, the US is one of the largest consumers of nuclear energy in the world, despite existing in an adverse public policy environment. It may come as a surprise to many people that the US has a much higher consumption versus production deficit in nuclear power than it does compared to its petroleum deficit, in proportion to its production levels.

The US produces a modest amount hydroelectric power. However, the per capita amount is not large, and production tends to be concentrated in the western part of the US. Nevertheless, hydroelectric power comprises about 70% of renewable energy production in the US. Biomass and geothermal sources comprise roughly 20% of other renewable energy sources. Wind and solar are quite minor sources.

The mid-term energy future of the US will be coal and nuclear power, along with moderate amounts of petroleum and natural gas. The US has coal reserves to last for over 200 years at present levels of production and consumption. The United States has large uranium reserves, and the pressure to increase nuclear energy production, specifically uranium mining, will be tremendous within the next few decades.

The long-term, sustainable energy future for the US will be hydroelectric in the west and biomass in the north central and northeast. Wind will be extremely important in low-population density areas in the north, and solar will be important in the south and especially the southwest, and significant in most other parts except Alaska.

UNITED STATES

(All figures are in terms of years unless otherwise noted)

Production and Consumption (National Totals)

Energy Source	Production (PJ)	Consumption (PJ)	Surplus (PJ)
Petroleum	18,700.00	42,600.00	-27,500.00
Natural Gas	20,100.00	23,600.00	-3,560.00
Coal	25,300.00	24,300.00	1,030.00
Nuclear	1,180.00	8,610.00	-7,430.00
Hydroelectric	2,850.00	2,850.00	0.00
Total	68,100.00	106,000.00	-37,500.00

Production and Consumption (Per Capita)

Energy Source	Production (kwh)	Consumption (kwh)	Surplus (kwh)
Petroleum	17,300.00	42,600.00	-25,400.00
Natural Gas	18,500.00	21,800.00	-3,290.00
Coal	23,400.00	22,400.00	952.00
Nuclear	1,090.00	7,940.00	-6,850.00
Hydroelectric	2,630.00	2,630.00	0.00
Total	62,900.00	97,400.00	-34,600.00

Reserves and Projections (National Totals)

Energy Source	Reserves (PJ)	Production (PJ)	Years
Petroleum	183,000.00	18,700.00	9.79
Natural Gas	221,000.00	20,100.00	11.00
Coal	5,840,000.00	25,300.00	230.00
Nuclear	257,000.00	1,180.00	217.00
Total	6,500,000.00	65,300.00	

For sources and notes, please see Section 3.

4.2.4 Region Discussion: NAFTA

With an overall highly developed economy and a total population of nearly one half billion people, the NAFTA region is one of the world's largest consumers of energy, but it also has moderately large energy resources. The NAFTA region appears to have a nearly balanced production and consumption of energy sources other than petroleum. Yet, this illusion masks significant flows of energy between NAFTA members and external countries. There are significant flows of electric energy among members.

The NAFTA region is a large net importer of petroleum. In fact the largest component of its energy consumption (in terms of Joules) is petroleum. NAFTA is a small importer of natural gas, despite its large production. The region's reserves of petroleum and natural gas are insufficient to meet current levels of demand (at current prices) for much more than a decade. The NAFTA region has tremendous reserves of coal, both overall and in each of its member states.

NAFTA, as a region, both imports and exports large quantities of uranium fuel. NAFTA possesses large reserves of uranium in terms of present production.

NAFTA, overall, only has modest hydroelectric power generation, and a relatively tiny amount of other renewable energy production.

The question for the NAFTA energy future is how much will its members share energy resources with each other, which will be a function of how significant the benefits of cooperation will be. If solar energy becomes extremely inexpensive, the south half will become the long-term energy "bread-basket" of North America. If renewable energy production remains identical to hydroelectric power generation, the northern half may be hold the title. Perhaps next to the cost of solar energy, the ability to produce biofuels in both an economic and environmentally friendly manner is the next biggest factor for the NAFTA region's long-term, sustainable energy future. North America has significant agricultural resources. The energy future bodes well is those resources can be maintained and in part used for energy, but if NAFTA population growth continues, biomass will not be sufficient.

[Please continue to the next page.]

NAFTA

(All figures are in terms of years unless otherwise noted)

Production and Consumption (NAFTA Region Totals)

Energy Source	Production (PJ)	Consumption (PJ)	Surplus (PJ)
Petroleum	34,300.00	55,700.00	-21,400.00
Natural Gas	29,000.00	29,300.00	-313.00
Coal	27,100.00	26,100.00	1,010.00
Nuclear	9,880.00	9,760.00	118.00
Hydroelectric	6,940.00	6,940.00	0.00
Total	107,000.00	128,000.00	-20,600.00

Production and Consumption (Per Capita)

Energy Source	Production (kwh)	Consumption (kwh)	Surplus (kwh)
Petroleum	21,700.00	35,200.00	-13,500.00
Natural Gas	18,300.00	18,500.00	-198.00
Coal	17,200.00	16,500.00	638.00
Nuclear	6,240.00	6,170.00	74.60
Hydroelectric	4,390.00	4,390.00	0.00
Total	67,800.00	80,800.00	-13,000.00

Reserves and Projections (NAFTA Region Totals)

Energy Source	Reserves (PJ)	Production (PJ)	Years
Petroleum	368,000.00	34,300.00	10.70
Natural Gas	298,000.00	29,000.00	10.30
Coal	6,020,000.00	27,100.00	222.00
Nuclear	590,000.00	9,880.00	59.70
Total	7,280,000.00	100,000.00	

For sources and notes, please see Section 3.

4.3 Energy sources

Petroleum, natural gas and coal are by far the largest sources of current production and consumption. Both Canada and the US consume an important amount of nuclear energy for electricity production. Hydroelectric production is important for the United States and especially Canada. Other renewable energy sources are still minor but growing.

4.3.1 Petroleum

Wars are fought over petroleum. That alone is sufficient to express the unique importance of petroleum as an energy source.

Petroleum is a liquid mixture of complex organic compounds that require about 100 million years to form in the right mixture of ecology, heat and pressure. Petroleum is the chief fuel used for transportation, as well as to power farming equipment required for food supplies. As its price increases, its use to generate electricity is waning, but steady population growth and economic development has resulted in skyrocketing demand. Petroleum supplies are also highly sensitive to adverse weather conditions such as hurricanes in the Gulf of Mexico.

Petroleum is expensive to discover and often expensive to extract. The chief product of petroleum is gasoline that requires further expense and infrastructure to produce. Consequently, the petroleum business is a risky one with cycles of boom and bust. It is difficult to increase production on short notice, although it can be decreased relatively quickly and easily, as major exporters have discovered to the detriment of low prices. Petroleum produces greenhouse gasses and its consumption contributes to greenhouse gasses. Even so, its consumption appears to be increasing.

The United States has the largest reserves of petroleum, unless optimistic estimates are used for Canadian oil sands. In any event, both Canada and Mexico have large reserves as well. The US is a large producer of petroleum, but high consumption both on a total and per capita basis make the United States a large net importer of petroleum. In fact, US consumption is so large that it renders the entire NAFTA region as a net importer of petroleum. Unfortunately, remaining reserves are often problematic to exploit. Oil sands require a tremendous amount of energy to utilize. Some reserves are in environmentally sensitive areas. Other reserves are underneath kilometers or miles of ocean.

4.3.2 Natural Gas

Natural gas is a gaseous mixture of relatively simple hydrocarbons. It is often found with petroleum, but there are gas fields separate from oil fields. Equivalent substances can be easily produced from decaying organic matter, but most natural gas is extracted from geologic sources. Natural gas is used to heat homes. It is used for electricity generation, but typically only during peak periods due to its expense. Natural gas is increasingly used as a relatively clean fuel to power vehicles. Consuming natural gas increases greenhouse gasses, but it produces less of other types of pollution.

Canada produces enough natural gas to both heat its homes and export a substantial amount. The United States is a large producer of natural gas but also a large net importer. Mexico is a minor importer.

4.3.3 Coal

Coal has been called "King Coal." Although coal has perhaps lost a few jewels in its crown to petroleum, coal is still king in terms of reserves.

Coal is decayed plant matter that has been crushed under high pressure over a period of millions of years. It comes in several distinct forms characterized by hardness. Sulfur content is also an importance factor that differentiates different types of coal. Coal is presently used chiefly to generate electricity. It used to be extensively utilized for transportation when locomotives ran on coal and before the gasoline-powered internal combustion engine had been invented. Coal consumption is not only a major contributor to greenhouse gasses, but it is also a major producer of environmentally harmful acid rain. That said, consuming harder coals tends to produce less sulfur than softer ones. Coal can be converted into automobile fuel, but this process is expensive. Efforts are being made to utilize coal energy without releasing its carbon.

At the present rate of production, the US has sufficient reserves of coal to last for over 200 years, while Canadian and Mexican supplies would last over 100 years. Coal exploitation appears to be less economically risky than petroleum exploitation. On the other hand, coal mining presents other dangers. Mining coal using traditional shaft methods is dangerous to miners. Using newer open pit methods can cause tremendous environmental destruction. Some of this environmental destruction can be mitigated, but the extent to which this is possible is disputed.

4.3.4 Nuclear

Nuclear power is the lowest cost way to produce electric power, but it is also the most immediately dangerous source. Electric power is produced when enriched uranium fuel is placed into a nuclear reactor and breaks apart into lighter elements. The nuclear fission generates heat that produces steam that in turn drives an electric generator. In Ideal circumstances, nuclear reactors are quite safe. If something dreadful goes wrong, they can be deadly to the surrounding community. Serious accidents are scarce though not unheard of. Nuclear power has been banned in many countries, but since it directly creates no greenhouse gasses, there is increased interest in nuclear power. Mining uranium can release radiation into the environment, which is no small matter. Consuming nuclear power presently creates large amounts of dangerous radioactive waste, but breeder reactors could be used to reduce the amount of this waste.

Both Canada and the United States have large reserves of uranium. Mexico has small reserves of uranium that are not presently economical to exploit. Canada produces and exports large amounts of uranium fuel, while the US is a small producer of uranium fuel but a large consumer.

4.3.5 Hydroelectric

In former times, "water power" was utilized to mill grain or weave textiles, but today it is used to produce inexpensive electric power. Hydroelectric power is generated when water, caught up in a dam reservoir, releases its gravitational potential energy when allowed to flow though an outlet containing an electric generator.

Hydroelectric power produces no direct greenhouse gasses, but it can lead to other environmental problems such as preventing fish from swimming up river to reproduce. Where fishing is an important industry, this can have serious economic repercussions as well. It can be difficult to increase hydroelectric power production, since many of the appropriate locations are already being utilized. Hydroelectric power is considered to be a renewable energy source, but since the required dams tend to silt up rivers, long-term use may require a continuous dredging and reconstruction program. Hydroelectric power is highly dependent upon rainfall. Global climate change could tremendously affect the production of hydroelectric energy in the future.

Canada produces most of its electric power by hydroelectric means (the remainder being produced chiefly by nuclear power). The United States produces less hydroelectric energy than Canada, but hydroelectric power is a substantial source in the western US.

4.4.6 Non-hydroelectric Renewable Sources

Non-hydroelectric renewable energy sources chiefly include biomass, geothermal, wind and solar power, but at present hydroelectric power generation is by far the majority of present renewable energy production.

Current renewable energy sources comprise only about 7% of total energy consumption. Hydroelectric power is the chief component, with wood fuels second, and geothermal sources third. Other forms of biomass are being increasingly used, and efforts are being made to use more biofuels whose production is not environmentally harmful. Solar and wind power only constitute a few percent or less of total energy production. Efforts are being made to greatly reduce the cost of solar energy.

Arguably, consumption should equal production, since long-term storage of power is not particularly economical. The figures suggest this is not the case. Reasons for the differences could be that not all of those figures are for truly renewable sources, or that some energy from those sources is produced in one country and consumed in another.

4.4 Country comparisons and energy flows

The United States is both NAFTA's largest producer *and* consumer of energy in terms of its national total. Canada is booth NAFTA's largest producer *and* consumer of energy in per capita terms. Canada falls second for both absolute production and consumption. Mexico has the lowest per capita energy consumption, but due to its significantly large population, it is a major user of energy.

Canada is largely self-sufficient in terms of supplying its own energy consumption. It exports energy to the United States particularly in the form of electricity, although this is becoming harder to measure. Canada exports oil, natural as and uranium. The United States exports small amounts of energy, but it is by far a net importer especially of petroleum and uranium. Despite the US military involvement, the Middle East provides only a small amount US oil supply. Like Canada, Mexico is also fairly self-sufficient in terms of supplying its own energy consumption. Mexico is an important exporter of oil to the US

4.5 Individual energy consumption

It is possible to calculate mean individual (per capita) energy usage. This is shown in Table 9. For example, if a hair dryer uses 1 kilowatt hour *per hour*, then it would be as if a typical Canadian was surrounded by 14 running hair dryers, a Mexican by 2 running hair dryers, and a person in the US by running 11 hair dryers. That's a lot of hot air!

Yet, these figures encompass total energy usage per capita per hour. Much of that energy is used to extract petroleum and coal, maintain dams, build and operate generation plants, power trucks, operate factories and light up stores and offices. Etc. By the time that energy finally reaches to consumer, it might be reduced by a largely unsubstantiated estimate of 80%. In that case, for example, the typical Canadian might only be able to continuously run 3 hairdryers. Discounting these figures by 80%, it would be interesting for consumers to compare their usage.

Another question is how many hairdryers could a person continuously run if one was limited to existing renewable energy sources (including hydroelectric)? Table 10 attempts to answer that question for the case of a person in the U.S. (assuming no Canadian hydroelectric energy is used). Such a person could in theory run a whole hair dryer on renewable energy sources only, but since it could take up to 80% of that power to deliver the remaining 20%, actually only about 1/5 of a hair dryer could be operated. Or make sure you can dry your hair in 12 minutes!

In contrast, Table 11 shows that a Canadian could continuously run nearly 4 hairdryers off of just hydroelectric energy alone (if Canada kept it all to itself). Discounted by 80%, the typical Canadian could still bask in 45 minutes of hair dryer time per hour at home from just hydroelectric energy.

[Please continue to the next page.]

Table 9: Per Capita Mean Energy Consumption Per Unit of Time

Country	Per Year (kwh)	Per Day (kwh)	Per Hour (kwh)
Canada	128,000.00	351.00	14.60
Mexico	19,500.00	53.30	2.22
USA	97,400.00	267.00	11.10
NAFTA*	80,800.00	221.00	9.22

*Note that the NAFTA figure here is a mean weighted by capita, not a total.

Table 10: USA Per Capita Mean Energy Consumption Per Unit of Time, Renewable Only

Country	Per Year (kwh)	Per Day (kwh)	Per Hour (kwh)
USA	9,214.00	25.30	1.05

Table 11: Canada Per Capita Mean Energy Consumption Per Unit of Time, Hydro Only

Country	Per Year (kwh)	Per Day (kwh)	Per Hour (kwh)
Canada	33,000.00	90.40	3.77

For sources and notes, please see Section 3.

4.6 Future technologies

New sources of energy and technologies include more cost-effective solar cells, the use of nano-technology to generate electricity, additional types of biofuels, and cleaner ways to use coal, just to name a few.

New technologies in cleaner and more efficient consumption of energy may be just as important as production technologies. Although technologies such as fuel cells and compressed air power are not energy sources themselves. They are means to utilize energy that can help minimize energy usage. Such technologies can have a significant effect on energy production and consumption patterns.

4.7 Potential trends

North America has significant coal reserves that could last for over two centuries. Reserves of other types of fuels may last less than one century. Currently, renewable sources (along with miscellaneous other sources) represent roughly 10% of present North American energy consumption. The lion's share of that figure is hydroelectric power.

Of major non-renewable energy sources, coal and nuclear present the longest projected reserves. To the extent that non-renewable energy sources continue to be used, it is expected that there will be a growing dependency from both of these sources. Of course, consuming coal creates greenhouse gasses that can contribute to global warming, so there could be environmental pressure to reduce coal production. Likewise, mining coal can cause severe environmental damage that will also put pressure to reduce its consumption.

Conversely, if the price of petroleum continues to increase at its present rate, it will become increasingly economical to literally trade coal for petroleum. If so, a key trend could be the conversion of coal energy into forms to supplement petroleum. This is already happening with the use of coal to generate electricity that powers increasingly popular electric vehicles. This literal trading could add pressure to increase consumption of coal. Then the coal reserves could last for a shorter period than estimated.

There are several other potential long-term trends to be considered. For example, an increasing export of energy resources to rapidly developing countries such as China and India, would likewise shorten the estimated period that coal reserves could last.

Promising new technologies could provide additional sources of renewable energy. Less expensive photovoltaic cells could substantially reduce dependency on other fuels for the generation of electric power.

5. AREAS OF POTENTIAL FUTURE STUDY

This study focuses on quantitative, physical aspects of energy production and consumption in North America. That focus spells out what areas this study has neglected that could be areas for potential future study.

A more in-depth examination of qualitative aspects of sources would provide more meaning and perspective to the numbers. Further, in today's global economic environment, North America is not an island. Supply and demand by countries outside of North America should be considered in order to provide a more complete picture.

Economic and financial perspectives also play an important role in North America's energy picture. One potential area of work in the future that would fit in well with this study's approach would be to consider gross domestic product (GDP) in terms of gross domestic energy consumption (GDEC). For example, what is a country's GDEC/GDP? What is the significance of this figure? Would that be a way to access a carbon or energy tax, particularly in Value Added Tax (VAT) countries?

Environmental concerns should be more fully examined. While this study provides figures that indicate how much heat is released into the environment, there are no figures for carbon dioxide release or specific impacts upon global climate change. The environmental impacts of extracting non-renewable energy sources as well producing renewable sources would be vitally important areas of study.

Yet the core area leaves room for improvement and additional work as well. For example, another area of future work would be to provide a more critical analysis of production, consumption and reserve figures. Petroleum reserves and nuclear power production are two areas for further consideration. Projections could be improved by providing figures over a range of years instead of a static single figure. Using an exponential decay method for projecting energy production may be appropriate.

Finally, other areas of future improvement could involve the drafting of the report itself. Language could be tightened up. Although this report is fully referenced, individual statements are not well-cited. This report could also benefit by the addition of graphics to better illustrate the quantitative results.

REFERENCES

B.B.C. News, *Country profile: Canada*. October 23, 2007.

B.B.C. News, *Country profile: Mexico*. December 6, 2007.

Deffayse, Kenneth S., *Hubbert's Peak: The Impending World Oil Shortage*. Princeton: Princeton University Press, 2001.

Ciotola, Mark, *Global Oil Geology and Economics.* San Francisco: 2005.

Eisberg, Robert and Robert Resnick, *Quantum Physics, 2nd Ed.* New York: John Wiley & Sons, Inc., 1985.

Energy Information Administration, *Table 1: U.S. Energy Consumption by Energy Source, 2001-2005.* 2007.

Energy Information Administration, *Table 17: Recoverable Coal Reserves and Average Recovery Percentage at Producing U.S. Mines by Mine Production Range and Mine.* 2007.

Energy Information Administration, *Table E.5 World Net Hydroelectric Power Consumption (Btu), 1980-2005.* September 11, 2007.

Energy Information Administration, *ENERGY INFOCARD - United States.* Viewed 02/03/2008 03:55 AM, http://www.eia.doe.gov/kids/infocardnew.html.

Energy Information Administration, *Domestic Uranium Production Report - Quarterly.* October 30, 2007.

Energy Information Administration, *International Energy Annual 2005, Table 82.* June 21, 2007.

Energy Information Administration, *International Energy Annual 2005, Table E1.* September 21, 2007.

Energy Information Administration, *International Energy Annual 2005, Table E.6 World Net Nuclear Electric Power Consumption (Btu), 1980-2005.* September 5, 2007

Energy Information Administration, *International Energy Annual 2005, Table F.1: World Primary Energy Production (Btu), 1980-2005.* September 20, 2007.

Energy Information Administration, *International Energy Annual 2005, Table F.6 World Net Hydroelectric Power Generation (Btu), 1980-2005.* September 11, 2007.

Energy Information Administration, *World Proved Reserves of Oil and Natural Gas, Most Recent Estimates.* January 9, 2007.

Energy Information Administration, *World Coal Consumption, Most Recent Annual Estimates, 1980-2006.* September 10, 2007

Energy Information Administration, *World Coal Production, Most Recent Annual Estimates, 1980-2006.* September 7, 2007.

Energy Information Administration, *World Dry Natural Gas Consumption, Most Recent Annual Estimates, 1980-2006*. December 21, 2007.

Energy Information Administration, *World Dry Natural Gas Production, Most Recent Annual Estimates, 1980-2006*. September 7, 2007.

Energy Information Administration, *World Petroleum Consumption, Most Recent Annual Estimates, 1980-2006*. January 14, 2008.

Energy Information Administration, *World Production of Crude Oil, NGPL, and Other Liquids, and Refinery Processing Gain, Most Recent Annual Estimates, 1980-2006*. August 24, 2007.

U.S. Census Bureau News, *Census Bureau Projects Population of 303.1 Million*. December 27, 2007.

Uranium Information Centre, Ltd. (Australia), *Canada's Uranium Production & Nuclear Power*, February 2008. http://www.uic.com.au/nip03.htm.